Flutterbye Flits, Floats, and Flies

**Written and Illustrated
by
Johnnie W. Lewis**

Marietta, GA USA

In this book, the third in the *Flutterbye, the Butterfly* series, Flutterbye's granddaughter has to learn to deal with the WEATHER, something her mother never told her about! Weather and climate affect all the animals in her neighborhood and they all have to learn how to deal with it!

Lewis, Johnnie W.
    flutterbye flits, floats, and flies/Johnnie W. Lewis

ISBN 978-0-9762559-3-2
Copyright © 2014 by Johnnie W. Lewis

For information regarding permission, write to:
Franklin Wright Enterprises
1860 Sandy Plains Rd.
Suite 204-150
Marietta, Georgia 30066
info@acloudproductions.com / www.acloudproductions.com

Text copyright © 2014 by Johnnie Wright Lewis
Illustrations copyright © 2013-2014 by Johnnie Wright Lewis
Cover art and design by Johnnie Wright Lewis
Published by Franklin Wright Enterprises.

Printed in U.S.A

Suggestions for Teachers/Parents

1. This book is designed to be READ by a teacher or parent to children or students ages 3-11, to allow for questions and answers. Some children, about 8 or older, can handle the verbiage, but children 12 and older should be advanced enough to read and understand everything in this book, including the "Teacher/Parent Notes" in the back of the book. **Media Specialists:** when checking out this book to Middle School students, point out the supplemental reading in the "Teacher/Parent Notes" section, for enrichment of the text. This book is the third in a series called the *Flutterbye, the Butterfly* series.

2. The section called "Teacher/Parent Notes" in the back of this book is designed to help augment the text of this book. The notes are divided by page applicability, e.g., notes at "Enrichment for page 1," should be read by you, the teacher/parent, before reading the text on page 1, and the information dispensed as you deem appropriate, based on the ages of your audience.

3. Almost all children love to be read to, and shown the pictures from that reading, though not all have the same ability to sit still long enough for the reading of this entire book. Attention spans of children differ, (1)depending on their ages, (2)the time of day/evening the book is read, (3)activities that are planned before or after reading time, and (4)the time of school year that the book is read (is it daylight or dark outside?). Suggested length of reading times include added discussion time to allow for questions and enrichment.
    a. Ages 3-7 -- 20 minutes.
    b. Ages 8-11 -- 30 minutes.
    c. Ages 12 and up -- the entire book.

4. Plan enrichment activities around food chains, or prey and predators in the discussion of "life circles." In the case of "Flutterbye," she is a second generation Monarch butterfly, which means she is born in April/May, and will not migrate as her grandparents did. Second Generation Monarchs will live for about four to seven weeks after she emerges from her chrysalis, but she will not migrate as her grandparents did to the southwestern U.S., Mexico, or Florida. Although the preceding books in this series, *Flutterbye, the Butterfly* and *Flutterbye's Flying Friends* would be appropriate for Spring application or "new life" time frame, this book is more appropriate for late Spring or early Summer application as this Second Generation Flutterbye is born, lives and dies in the area of her birth and, along with NUMEROUS animals and birds in her area, they all have to deal with the weather and climate of their habitat during their lives.

5. Plan a series of discussions on toxicity. Touch on subjects such as environmental impact of man-made toxic materials and wastes as opposed to wastes and materials that are toxic to humanity and animals,

but are created by Nature. Such as a Monarch Butterfly and its caterpillars.

6.   Plan a teaching unit on defense mechanisms of animals and plants, poison and poisonous creatures as opposed to venom and venomous creatures. Show the differences between those plants, insects, and animals that are poisonous to humans and other animals (to eat, to the touch, etc.) as opposed to those that inject poison into a human or another animal by means of a bite or a stinger. Ex.:
   • A Monarch butterfly is poisonous or toxic to most other creatures if eaten, but it won't hurt you if it lands on you. Its *body* is only toxic to you if its body gets inside yours (or an animal's body). Its defense mechanism is in the hope that you learned from eating the LAST Monarch butterfly, and that you won't want to eat this one.
   • A Rattlesnake injects venom, from poison sacs near its mouth, into its victims if it bites the victim, but its body won't hurt your skin if it glides across it. And eating rattlesnake meat won't hurt you because its flesh does not contain poison, just the poison sacs do. Defense: if you were poisoned by the last bite (or hear its rattles!), you will steer clear of the snake this time.
   • Some plants have stinging nettles, which are toxic if embedded in the skin, designed to keep people and other animals away from it. But eating the flesh of that plant won't kill you. Defense: you will walk around the plant the next time you see one!

7.   Plan art, science, research, group, or field trip activities around this book, by teaching a unit on flying creatures (birds and insects), and focusing on:
   a. The importance of insects that pollinate plants.
   b. The differences in plants and their usefulness to an area (Hint: Monarchs like to lay their eggs on milkweed plants [so their baby caterpillars will have something to eat immediately after hatching], which is a weed and not wanted in most cultivated or planned landscape areas. Without milkweed plants, will you get Monarchs to lay eggs?).
   c. Fragility of insects versus their strengths.
   d. Usefulness of flying insects versus crawling insects (and arachnids) versus burrowing insects (and arachnids).
   e. The beauty and symmetry of butterflies' and birds' and dragonflies' wings in flight.
   f. How wind and water damages different habitats around the country and the world, e.g., hurricane versus tornado damage, water versus wind damage, and how those differences effect the flying creatures.
   g. Draw, color, and cut-out different parts of the water cycle and place on mobiles or attach to the walls around the classroom. This will provoke questions and discussions during the day.
   h. A series of essays, for older students, discussing flying creatures and their evolution, psychology, imagery, etc. See the author's books called *The Five Finger Paragraph* series on Amazon.com, for more information on implementation of this suggestion.

DEDICATION

This book, and all such books by me, are dedicated to the child in all of us, who love nature, who love life, and who love butterflies.

Special thanks go to:

★My Better Half, Jimmy Lewis, for his undying love, support, kindness and a second pair of hands!
★My children, Tash Lewis White and Trevor MacKenzie Lewis. I would never have been inspired to write for children if I had not had you. Thanks for your tolerance of your errant mother!
★My "children-in-law," Troy White and Emily Xander Lewis. Though you were not born to me, I love as though you were my own children. Thanks for putting up with me!
★My grandchildren, Parker and Avery White. You are the lights in Mammy's eyes. Thanks for being my audience, my sounding boards, and my models (page 22). I love you!
★All of my "photographers," especially John Humphreys, Sandi Spires Nobles, and Jerry Battle. Your work is impossible for me to duplicate with colored pencils or paints!
★Children everywhere who want to learn about nature in our lives. Chase your dreams as you do butterflies, birds, and bees!

INTRODUCTION

There are four generations of Monarch Butterflies in North America every year. Three generations of Monarchs live and die in the areas of their births, but the Monarchs from the fourth generation of Monarchs each year will fly to more southern climates, to survive the winter, then return to their home areas to lay eggs before dying.

The "Flutterbye, the Butterfly" you will meet in this book is a **Second Generation Monarch butterfly**, born in April or May of the year. She is a daughter of the First Generation Monarch Butterfly from the second book in this series, *Flutterbye's Flying Friends*. *THAT* Flutterbye lived for four to seven weeks before laying eggs and dying.

THIS Flutterbye will only live for four to seven weeks after she emerges from her chrysalis, living and dying in a small area, of usually two to three acres or less. She will eat (or be eaten!) in the same area where she was born. Her children (Third Generation Monarchs) will do the same, live in that same area. But her grandchildren will be like her grandmother, Fourth Generation butterflies. And THEY will fly away to the South and come back to lay their eggs before they die.

Flutterbye's friends, the creatures in her **habitat***, are mostly flyers, but she and they must ALL learn to live with the weather and climate in their area. (NOTE: words in **bold** with an asterisk * are defined in the Glossary.)

And now, for the sights and experiences of Flutterbye's life, during the Spring and Summer of the year!

Flutterbye did not always look as she does now. She started out, as most other flyers in her **habitat*** do, as an egg. But when she hatched out of that egg, she was a caterpillar. Actually, she was more like a stomach with legs, because she ate the leaf she was born on and the one next to that. And the one above it and below it. And every leaf on that plant plus two more, growing twenty times her birth size. In one day!! Then, the next day she had to start all over again!

As she eats ALL the milkweed plants in her nursery, she breaks out of her skin (because she gets too big for her britches!) about four times (that's called **molting***) before she finally attaches herself to a leaf or branch, when her skin gets hard for the last time (and becomes her **chrysalis***), and she goes to sleep. During that "sleep," she will turn into a beautiful Monarch butterfly!

As a Monarch Caterpillar egg on a leaf, this is how Flutterbye, the Butterfly began her life!

Flutterbye was a beautiful caterpillar…

…and she lived in a beautiful chrysalis.

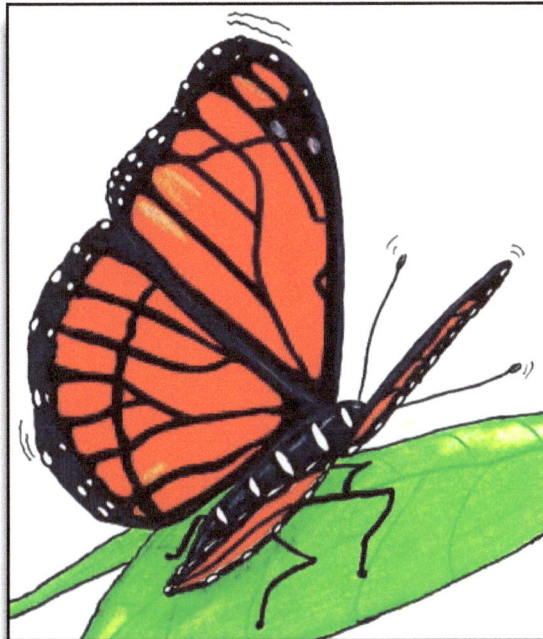

And THIS is what Flutterbye, the Butterfly looks like now, after she breaks out of her chrysalis!

1

Flutterbye, the Butterfly looks very much like her mother and grandmother. Her father looked a little different than they did, because he had a couple of black spots on his hind wings that her mother did not have. But, basically, most Monarch butterflies look alike. They all start life as eggs, laid on milkweed plants, because that is their food. All they have to do is crawl out of that egg and eat the plant where they were born!

But milkweed plants are poisonous to most animals and birds! And, even though eating milkweed does not bother Monarch caterpillars, it makes their bodies **toxic*** to other animals that try to eat them. Perfect defense for fat little caterpillars! That keeps most larger animals and birds from wanting to eat Monarch caterpillars, so that should make life easy for them, right? NO!! There is ONE thing that affects all plants and animals. Can you guess what that is?

Male Monarch (see the black spots in the orange on his hind wings?)

2

The **WEATHER***! All plants, animals, birds, and insects (and humans!) are affected by the weather! Is there a lot of rain? Or too much frost? Or how about not enough rain or sun? Or what happens to Flutterbye if there is too much COLD or HEAT? Or fog? Those are questions we are going to look at today. We will see how **weather*** works and how it works for or against Flutterbye and her friends!

Come to think of it, what is **weather***? And what is a word that is closely related to weather, called **climate***? **Weather*** is a combination of **factors*** that make up what it is going to feel like to you when you are out-of-doors. Think about it. What is the last thing you do before you leave to go to school? You check outside (or someone older checks for you) to see if it is raining, hot, cold, windy, or snowing, so you will know what clothes to wear for the day and whether or not you need to take a coat, umbrella, hat, gloves, boots, or sandals. If it is going to be hot for most of the day, you want to wear shorts and sandals. If it might snow, you will want snow shoes and a coat and gloves. If it is going to rain most of the day, you want an umbrella or raincoat.

Flutterbye has to deal with the weather when she starts off her day, too! But she does not have coats or water boots or gloves or anything except…, herself to protect…, and from whatever is happening outside!

So, let's see what it is like in Flutterbye's world, as she and her friends take care of themselves in the weather each day!

The least amount of "weather" that Flutterbye needs to worry about is a calm day, with no wind, no rain, and not too much heat or cold. That's when she just "flits" around all day. But the Spring of any year is not always calm. A Spring day is usually cool in the morning, warm during the middle of the day, followed possibly by a storm, with thunder and lightning. Or a sudden snow storm, when all of the flowers are blooming. A Spring shower could be a **torrential*** downpour or just a few light drops that fall to "settle the dust."

But think about it. Flutterbye, the Butterfly **flits*** around, all day long, from one flower to the next flower to get her food. But can she FLY when it is raining? What happens to the **scales*** on her wings when it is raining? Or when it is too windy? Or there is a snowstorm?!? Flutterbye just has to keep moving!

Flutterbye's scales look very much like this picture. She can still fly if she loses a few individual scales, or even if she loses a lot in one section of a wing. But if she loses too many scales in one area or too many over all four of her wings, she can not fly very well. If she can not fly, she can not get to all the flowers to sip nectar. If she can not sip nectar for her food, what happens to her?

So, you can see how important the weather (wind and temperature) are to Flutterbye and her way of life. And to ALL of her insect and bird friends. Weather affects whether or not the early bird gets the worm or the bats find mosquitoes.

Butterfly scales

Here's a picture of the **water cycle***, which is one of the **factors*** that make up the weather. In between the stages of this cycle, Flutterbye has to live. She needs to know when a bad storm is coming and she may not have had enough nectar today. Or when the sun has not shown for days and she really *needs* some sunshine for more warmth and energy. Each of those kinds of weather affects whether or not Flutterbye gets to eat, or has a place to sleep.

Ever wonder how Flutterbye FEELS when those things happen? Or do you think she HAS feelings?

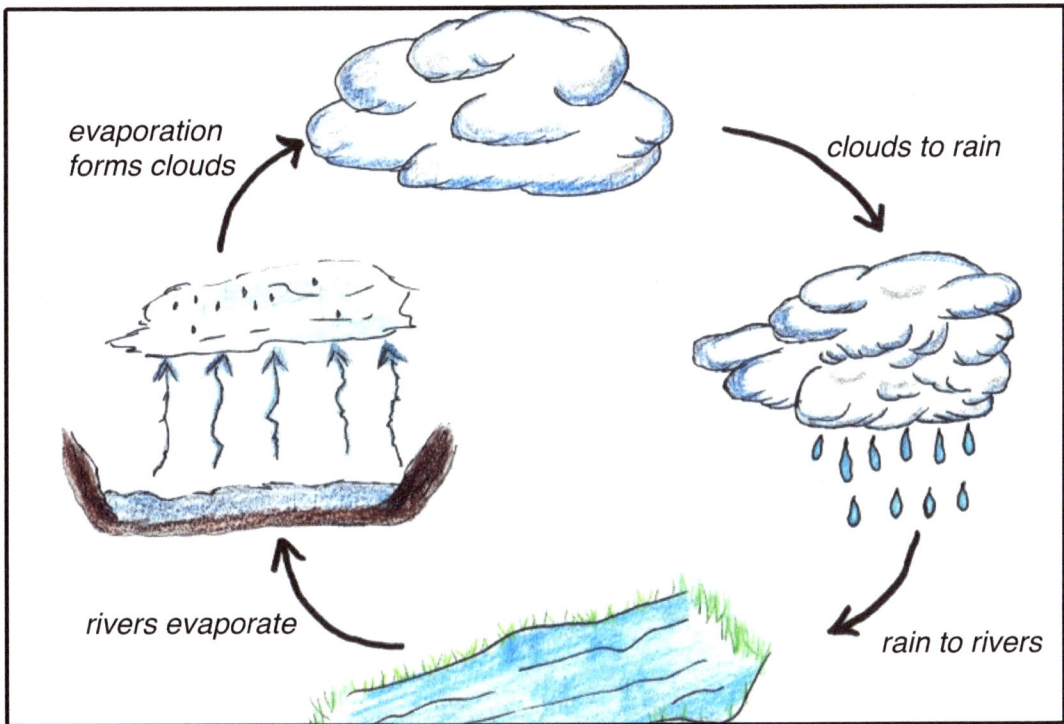

evaporation forms clouds

clouds to rain

rivers evaporate

rain to rivers

The Water Cycle

Wind is the one thing that Flutterbye **encounters*** almost every day. When there is no wind, she **flits*** or flies around. When she's flitting around, it looks like she is just bouncing around from one place to another, in the air, like her friend in this picture. It does not really look like she knows where she is going.

And, when there is no wind, she has to fly on her own. That is when she really needs to use her wings, to fly from flower to flower, using up a lot of her energy. That is when she is not getting any help from the wind!

8

But when there is a light breeze, Flutterbye gets to *float* around in the air. She really does not have to flap her wings much, so she uses less energy. She just floats, letting the wind carry her around. She only weighs as much as a small flower, so it does not take much wind to push her around.

But when the wind gets *REALLY* intense and moves very quickly, what do you think Flutterbye can do to protect herself?

There is not much she can do. Except hide! What do the birds do when the wind is very bad, such as during a **hurricane*** or a **tornado***?

Small creatures hide in bushes, under tree limbs, in old barns and buildings, almost anywhere they can get to that will shield them from the wind and rain. Even human beings are afraid of being out in extreme weather like severe wind and rain storms, so tiny creatures like insects have to find hiding places, and just wait until the bad weather goes away!

When tiny raindrops, such as mist or fog, fall on Flutterbye's wings, she can flap her wings **vigorously*** to knock most of the water, and its weight, off her wings, and continue to fly. But, since Flutterbye herself weighs so little, the added weight of a couple of drops of rainwater on her wings might make them too heavy for her to flap enough to keep her in the air!

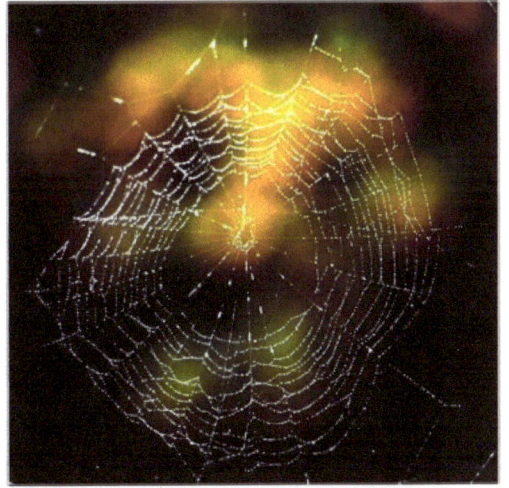

When hard rain falls and hits her wings, it can knock off some of her scales and even knock her to the ground, making it very difficult, or even impossible, for her to fly again. Ever. So, she has to find shelter when the rain starts, just to protect herself. The other small creatures usually seek the same kind of protection, too.

Just imagine what you would find under a bush or in hiding during a **tornado*** or **hurricane***!!

Another thing that happens when it rains is that the places Flutterbye and other insects go to get food are filled up with water! A tiny bit of the water will get down into the nectar in a flower and will **dilute*** it. When **nectar*** is diluted, it is not as "energy efficient" as it was before it was diluted. The creatures have to drink a LOT more liquid to get the same amount of energy that they were getting before the rain. It is like drinking a drink with ice in it, like a Coca-Cola or tea, in which the ice has melted and made the drink taste watery. Still a little sweet, but not as good as it was! After rain, these creatures have to drink twice as much to get even half the amount of energy they need to fly!

But Flutterbye, the Butterfly can drink upside down! So, when it rains and the flowers on TOP of the bush get too much water in them, Flutterbye can go to the flowers that are facing the ground and get her nectar. The nectar does not leak out of the flower, and since the rain could not get into the upside-down flower, the nectar is not diluted! Does that make you wonder if the other insects thought of that, too?!?

14

Something else Flutterbye probably never worries about, but that happens anyway, is the AMOUNT of rain that falls in her **habitat***. If there is too much rain during her short lifetime, the plants where she goes to get her nectar can get "**water-logged***" or even start growing molds on them, which make the plants sick. Too sick to produce good flowers, with nectar, for Flutterbye.

On the other hand, if there is not ENOUGH rain, what might happen to the plants and flowers? That's right! During a **drought*** period, the plants and flowers either do not grow as well or will die altogether. If ALL of the plants in her area die, how will the butterflies live? Will this friend of Flutterbye's get enough nectar to give her enough energy to fly?

Sometimes, people put out feeders for hummingbirds, and butterflies and other insects visit those feeders so they can get the nectar (sweetened water) that the people provide. Does that make you want to put out a feeder to help feed the flying creatures? If you live in a very dry **climate***, that's probably a very good idea!

Remember the storms that we talked about earlier, the thunderstorms? Most of the time during a thunderstorm, there is **lightning*** and **thunder*** that comes with the rain. Butterflies do not have ears so they can not hear the noises from thunder and lightning, but they can feel the **concussion*** caused by the booming of those noises in the air. This concussion might push Flutterbye around in the air! And if she were struck by lightning, she would be a Fried Flutterbye! So, when rain and other bad weather start, Flutterbye and her friends ALL run for cover!

Sometimes, **freaky*** things happen in Flutterbye's home area. One day it may warm and sunny during the late Spring, and suddenly the next day, there may be snow falling and ice forming. If enough snow falls and covers up all of Flutterbye's flowers, how will she get her nectar? Snow that falls in the Spring usually melts pretty quickly, so Flutterbye should be alright. And so should her friends.

18

But, if ice forms on all of Flutterbye's flowers, they will usually die. If the flowers die, what happens to Flutterbye? Will she die, too? Maybe. If the ice does not stay long, she might live. Most of the birds, and even some of the insects (except the mosquitoes), will **survive*** a "flash frost," but the air around them will still be very cold for a while.

We mentioned a word earlier that might be unfamiliar to you. **Climate***.
Weather is the daily events that happen around you. Rain. Hurricanes. Sunny and hot. Breezy or calm. Snow and ice. Those events and conditions happen each day or week.

But **climate*** is not a daily "thing." Climate is a long-term result of all of the days of weather. Climate is what happens in an area of the country over many YEARS of weather. Flutterbye likes to live in warm climates, that are not TOO hot and not too cold.

Which picture looks like what the climate is like where you live? Do you see anywhere that is similar to your home area? Could Flutterbye live where you live?

22

So. What have we learned here? First, we looked at Flutterbye being born and learning to live. Then, we found out that Flutterbye and ALL of her friends have to learn to live with the WEATHER! Just like WE do!

The weather and climate in an area can make life very pleasant for Flutterbye or rather **treacherous***, depending on the amount of wind and rain and snow and tornadoes. Or just the gentle breezes…

Now that we have learned all about areas where Flutterbye might live, we can better imagine what life must be like for Flutterbye and her friends!

GLOSSARY

chrysalis -- *noun* \'kri-sə-ləs\: (1)a moth or butterfly at the stage of growth when it is turning into an adult and is enclosed in a hard case; (2) a hard case that protects a moth or butterfly while it is turning into an adult.

climate -- *noun* \'klī-mət\: (1)a region with particular weather patterns or conditions; (2)the usual weather conditions in a particular place or region.

concussion -- *noun* \kən-'kə-shən\ (1)a stunning, damaging, or shattering effect from a hard blow; (2)a hard blow or collision.

dilute -- *transitive verb* \dī-'lüt, də-\(1)to make (a liquid) thinner or less strong by adding water or another liquid; (2)to lessen the strength of (something).

drought -- *noun* \'draút\ a long period of time during which there is very little or no rain.

encounters -- *verb* \in-'kaún-tər, en-\ (1)to have or experience (problems, difficulties, etc.); (2)to meet (someone) without expecting or intending to.

factors - - *noun* \'fak-tər\ (1)something that helps produce or influence a result; (2)one of the things that cause something to happen.

flits -- *intransitive verb* \'flit\ moves or flies quickly from one place or thing to another, seemingly without reason.

freaky -- *adjective* \'frē-kē\ strange or unusual.

habitat -- *noun* \'ha-bə-‚tat\ the place or type of place where a plant or animal naturally or normally lives or grows.

hurricane -- *noun* \'hər-ə-‚kān, -i-kən, 'hə-rə-, 'hə-ri-\ an extremely large, powerful, and destructive storm with very strong winds that occurs especially in the western part of the Atlantic Ocean.

lightning -- *noun* \'līt-niŋ\ the flashes of light that are produced in the sky during a storm.

molting -- *verb* \'mōlt*biology* : losing a covering of hair, feathers, etc., and replacing it with new growth in the same place.

nectar -- *noun* \'nek-tər\ a sweet liquid produced by plants and used by bees in making honey.

scales -- *noun* \'skāl\ small, flattened, rigid, and definitely circumscribed plates forming part of the external body covering, especially of a fish; small tiles that contain butterflies color.

survive -- *verb* \sər-'vīv\ to remain alive; to continue to live; to continue to exist.

thunder -- *noun* \'thən-dər\: the very loud sound that comes from the sky during a storm; the sound that follows a flash of lightning.

tornado -- *noun* \tȯr-'nā-(‚)dō\ a violent and destructive storm in which powerful winds move around a central point.

torrential -- *adjective* \tȯ-'ren(t)-shəl, tə-\ coming in a large, fast stream.

toxic -- *adjective* \'täk-sik\ (1)containing poisonous substances; (2)containing or being poisonous material, especially when capable of causing death or serious debilitation.

treacherous -- *adjective* \'tre-chə-rəs, 'trech-rəs\ (1)likely to betray trust; (2)providing insecure footing or support; (3)marked by hidden dangers, hazards, or perils.

vigorously -- *adverb* \'vi-g(ə-)rəs-lee\ (1)healthily and strongly; (2)done with great force and energy.

water cycle -- *noun* \ the series of conditions through which water naturally passes from water vapor in the air, to being deposited (by rain or snow) on earth's surface, and finally back into the air, especially as a result of evaporation.

water-logged -- *adjective* \-‚lȯgd, -‚lägd\ (1)so filled or soaked with water as to be heavy or hard to manage; (2)saturated with water.

weather -- *noun* \'we-<u>th</u>ər\ the state of the atmosphere with respect to heat or cold, wetness or dryness, calm or storm, clearness or cloudiness.

TEACHER/PARENT NOTES
(only those pages requiring "enrichment" or "enhancement" will have Enrichment NOTES)

Enrichment for pages 1

Build a diorama of a butterfly's life. Use photographs taken from various online sites marked "public domain," such as USDA or .gov sites.
========================

Enrichment for page 2

Look for some household items that will demonstrate what a pretend "toxic" (but not poisonous) reaction might look like, in the classroom or around the house. Try mixing a few items together to see what "blends together" or what causes "reactions" when stirred/mixed together. Here are some non-toxic ideas that will "react" to each other. Suggestions: (1)Put 2 teaspoons of baking soda in a glass and add a tablespoon of vinegar. The mixture will "boil" up. (2)In a bowl of salted water, sprinkle pepper on top of the water. Put dish soap on your fingertip and immerse your fingertip in one side of the bowl. Pepper will "run" to the other side of the bowl.

NOTE: PLEASE don't use gasoline or drain cleaner or other products that contain lye or caustic substances in your "experiments."
========================

Enrichment for page 3-4

Boil some water and let students see how the steam rises, before explaining the "water cycle." Add ice cubes to the water as it boils, to show how the rapid additions to the water cycle changes how it works. If water is boiling quickly enough, ice cubes should crack and pop.
========================

Enrichment for page 5

Put a student's hand inside a plastic bag and put the hand and bag in water, to demonstrate how the snow boots and coats protect from the weather. This is truly effective if you live in a warm climate, where the student(s) never or rarely see(s) snow.
========================

Enrichment for page 6

Make Butterfly Scales. Cut a piece (or two pieces) of notebook paper (so you'll have the lines on it to see) into 1" squares. Line up 8-10 squares, starting at the bottom, across the width of another piece of paper. Tape them down just along the top edge of the squares, so that the bottom edges are "flapping" free. For the next row, move up about 1/2" and line up another row of squares over the ones already taped down, so that the edges of each square of this new row are in the middle of the square below it. Tape down this row, just along the top edge of the squares.

Your rows will look like rows of bricks, but will be overlapping with a "free" edge, like roofing shingles. Continue to tape rows of overlapping, and off-set, squares until you have four or 5 rows. Even more effective if you turn the squares with the lines pointing up and down instead of side to side.

Now that you've made your "Butterfly Scales," allow student(s) to flip the under-page so that he can see the "scales" flap in the breeze. Or blow on them from the loose ends of the scales. Project this into an "imagine this" scene, where your squares of paper are Butterfly Scales. NOTE: Pick a species of butterfly and if you do this with smaller colored pieces of paper, you can imitate the pattern of scales on a particular butterfly's wing.
=========================

Enrichment for page 7

Build a diorama of a water cycle, using cotton balls for clouds.
=========================

Enrichment for page 8, 9, and 10

Cut out pictures of butterflies (or draw, cut-out and "build" butterflies) and attach a string to each butterfly's back. Tape the strings to a yardstick, so that butterflies are flying "free." Move the yardstick in different ways so that you can demonstrate the difference between flitting, flying and floating, explaining the effects of the wind in each case.
=========================

Enrichment for page 11

Weigh the "Butterfly Scales," from #6 above, on a postal or digital scale so you have a starting weight (something that will register the weight of very LIGHT weight items). Add a few drops of water to the scales, then re-weigh the "scales," to show how much heavier the scales would be on the butterfly with raindrops added.
=========================

Enrichment for page 12
Let student(s) make a list of the places (or bring in pictures of the places) that butterflies, bees, mosquitoes, grasshoppers, ants, etc. could hide during a wind or rainstorm and use that list in a spelling test or spelling bee.
==========================

Enrichment for page 13
Demonstrate the dilution of the nectar by pouring small amounts of Kool-aid, Coke, tea, etc. in small cups. Students drink the first "round" so they'll have the "normal" taste in their mouths. Now, pour more small amounts in the same cups, but add a 1/2 teaspoon or so of water to the sweet drink. NOW they will understand diluted nectar!
==========================

Enrichment for page 14
Don't think you want to try to demonstrate this one! You'd have Kool-aid, Coke, etc. all over the floor, as each student tries to drink while upside down! Just let them imagine the situation or "try this at home"!
==========================

Enrichment for page 15
In two separate containers, place or grow seedlings of some plant. When they're about 4" high, flood one with too much water (making sure the water can't escape). Add salt to the water, to demonstrate the effect of sea water being brought ashore in a hurricane, and allow to sit for a few days. Note how long it takes for the plant to start "dying" or changing colors.
==========================

Enrichment for page 16
Unless you have a drought in your area and can show examples of its effect, build your own "drought stricken" area. Place a plastic tray with mud or wet dirt under a desk lamp that stays turned on. Watch the effect of too much sun and not enough rain on the mud/dirt, noting how long it takes for the ground to "dry up" and crack.
==========================

Enrichment for page 17
Demonstrate "concussion" of the air by bringing in a drum or Oatmeal box or other box with a tight-fitting lid. Place box on one side, beating on the side of the empty box, like you're beating a marching bass drum. Place one hand on the opposite side of the "drum" from the side you are beating on and feel the concussion of the beats.

OR, to SHOW the concussion effect, inside the box, mount or hang a piece of paper so that one edge is "free" then seal the

Enrichment for page 17 (cont.)

box. You have 5 sides of the box showing (it is sitting on the 6th side), 4 "sides" and a top. Cut holes in two opposite sides of the box, then seal those holes over with clear plastic taped to the outside over the holes or with clear packaging tape. Anything will work, so long as students can see in the holes and the holes are sealed up around the edges.

Now. You've got the sealed box, with "viewing holes," and a piece of paper mounted or hanging inside the box. When you beat (hard) on the box on one end, the paper will move inside the box. If not, the box is not sealed up tightly enough. The movement of the paper demonstrates how the concussion of sound waves impacts objects (and people and animals).
==========================

Enrichment for page 18

Demonstrate making snowflakes. Fold a piece of paper in half across the width, then in thirds across the folded edge (see drawing on next page) so that you have one triangle. You will have six triangles in the folded paper. Cut through all layers of the paper in any design (since no two snowflakes are the same!), then open the papers and see the varied snowflakes! NOTE: Cutting off the cornered edges of the paper will make for a more realistic looking snowflake.

If you have older students (4-8), once the paper is folded in half, have them measure the angles with a protractor, dividing the halved paper into thirds of 60 degrees each, before cutting the "6-sided snowflake."
==========================

Enrichment for page 19

Spray a lot of water on a flower and place it in the freezer. When removed, note how much heavier the flower is and how long it remains viable after the ice melts.
==========================

Enrichment for page 20

Let students peruse the pictures of diverse climates and choose the picture(s) where they think they might like to live. Or where their grandparents live. Or a place they have visited.
==========================

How to make a paper snowflake.

1. fold page in half across the width

2. fold page in half to find center point on folded side

3. fold page in thirds (may have to guess)

4. fold page in thirds

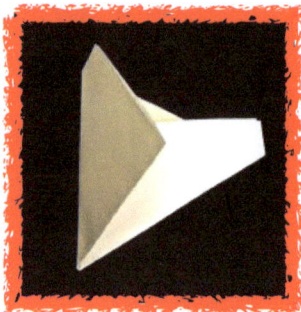

5. cut off excess paper

6. cut off and cut out everything that does *not* look like a snowflake!

10. your paper snowflake!

30

Acknowledgements

The following people were kind enough to submit their own photographs for this book. Photographs in the public domain are so noted. All illustrations are by Johnnie W. Lewis.

Jerry Battle: pages 3 (lower left), 11(geese and rose), 12 (upper left and right, lower right), 13 (bottom), 21 (upper right, alligators).

Allison Howard: pages 4 (rainbow and lower right), 17 (middle right), 21 (rainbow).

John Humphreys: pages 3 (balloons and lower right), 13 (middle left, lower right), 14 (middle left and middle right), 18 (upper left), 20 (lower right).

Theresa Lambe: page 17 (upper left).

Johnnie W. Lewis: pages 1, 6, 7, 21 (middle left [Las Vegas]), 22 (middle left), 23 (all except bottom), 30 (all).

Sandi's (Spires Nobles) Photography: pages 2 (upper and left Monarchs), 3 (upper left, middle left, and frost), 5, 8, 9, 11 (both spider webs), 12 (middle left and right, bottom), 13 (upper right), 14 (top), 15 (middle left), 18 (middle left, middle right, and bottom), 19 (all), 20 (sand dunes and crab), 21 (rocket).

Sharon Woodard Oliver: page 14 (bottom).

Edda Tucker Patterson: pages 20 (sea lions), 22 (raven), 24.

Tash Lewis White: pages 21 (upper left), 22 (upper right, lower right).

Debbie Jean Wright: pages 17 (upper right), 21 (white peacock).

Public Domain photos from USDA sites: pages 2 (male Monarch), 10 (all pictures), 13 (upper left), 15 (all except middle left), 16 (all), 17 (middle left [C. Clark], bottom), 18 (upper right), 22 (crane), 23 (bottom).

32

www.ingramcontent.com/pod-product-compliance
Lightning Source LLC
Chambersburg PA
CBHW041959100426
42813CB00019B/2936